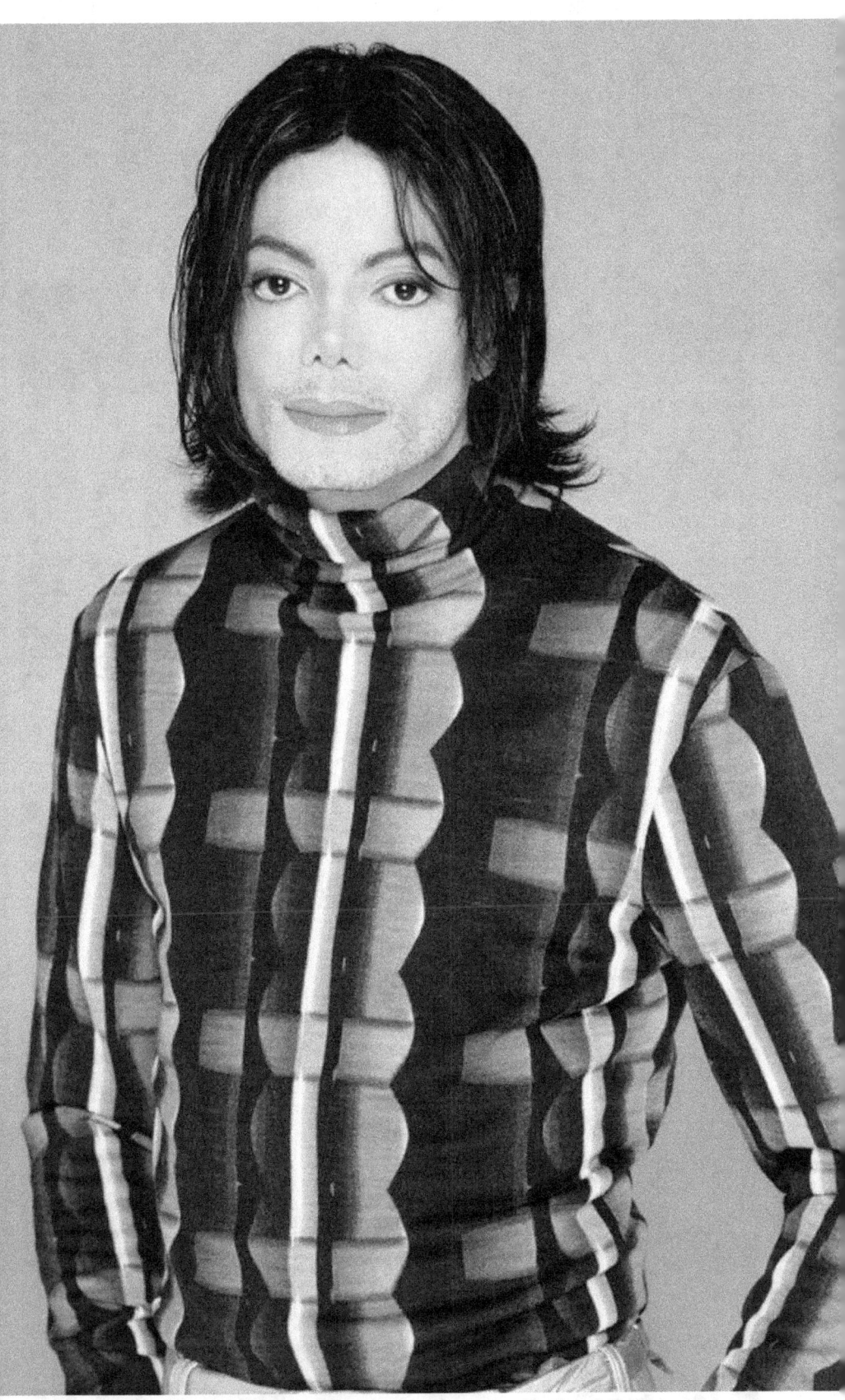

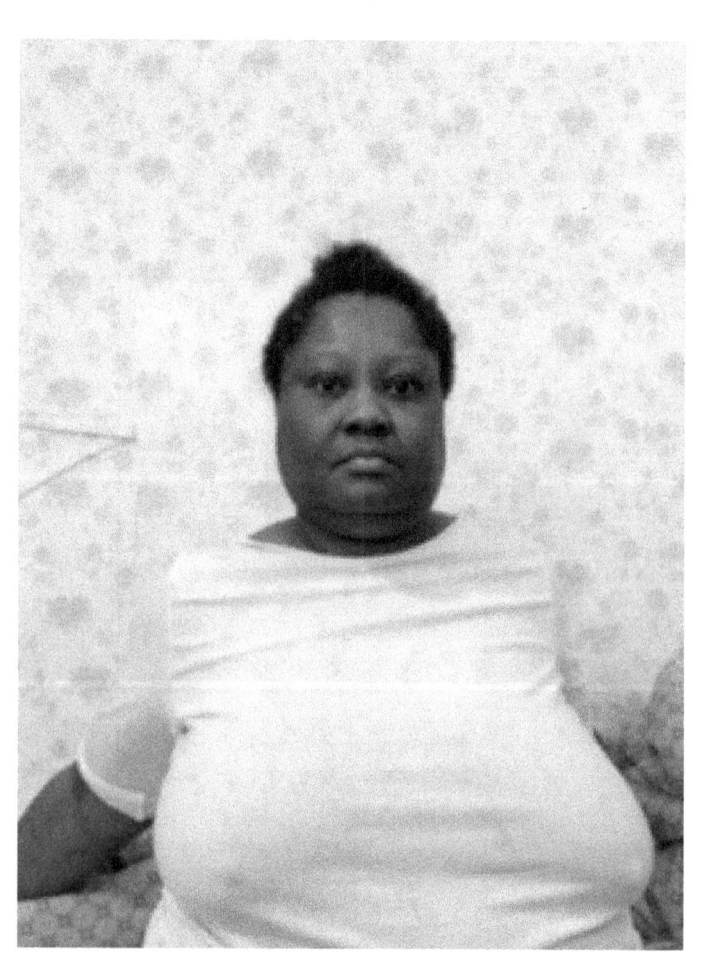

Your good leg is always the carrier of the burden of the bad (broken) leg

8:15pm: omega3 x4

9.08pm: menopace x1 + ashwag x1
+ uralds x1

9.51pm: silica x1 + amino15 cox1
+ garlic x5

10.32pm: Chromium picoinate x2
+ thydo collag x3 + Brewers yeast x2

11.16pm: Brewers yeast x4 + arginne x1
1tchy ankle bone (inner)

4.30am (Feb 23 2019) Brewers yeast x4
uva x1

4.45am. amino15 cox1
thydo collag x1 + chromium x1
+ arginne x1 + flaxseed x2 +
omega3 x2

Start
Pushed by a
Ghost

Excess uric acid in eyes
= uva ubsi + L-Arginine +
Evening Primrose oil

5:05am : even Prim x1 cloudy
left eye

5:38am : ashwag x1

9:52am : uva x1 + omega3 x2
+ arginine x1 + saw palm x1
+ Brewers yeast x4

10am : flaxseed x2

10:17am : chromium x1 +
amino 15000 x1
large bin out
11:16am : flaxseed x2 + uva x1

12:25pm : menopace x1 +
hydro collag x3

10.44pm: flaxseed x 2 + hydro collag x 2
mouth sore
11.13pm: omega3 x 3 + ura x 1 + ashwag x 1
+ arginine x 1 + face exfol. toptail

12.05am Feb 24 2019
fenugreek x 1 + l-arginine x 2

1.17am: mid back spinal attack +
flaxseed x 4

2.35am: ashwag x 1 + flaxseed x 4

2.57am: lysine x 1

7.46am: lysine x 1 + aminolsux x 1
+ hydro collag x 2 + ura x 1
+ l-arginine x 1

1:54pm: Brewers yeast x2 + tan x1 + chromium x1 + saw palm x1

3:47pm: Iron x1 + saw palm x1 + omega3 x4

4:07pm: arginine x1 + amino 15cap x1

5:52pm: pain in left leg + cranberry x6

8:12pm: flaxseed x3 left leg pain

8:25pm: ashwag x1

9:17pm: cranberry x5 + ura x1 + aminoison x1 + Brewers yeast x4 + chromium x1
mouth sore

Chromium mineralizes
the bone

8.14am: flaxseed x3

8.49am: ashwag x2

12.11pm: lyhex1 + hydro collag x2 + flaxseed x2 + ura x1 + arginine x1 + ashwag x1

1pm: washed (coconut) my hair + toptan + exfoliation + l-arginine x3 + fenugreek x2 + chromium + iron tonic

1.17pm: ashwag x3 + Saw Palmetto x1

1.50pm: aminos cux1

2.46pm: ura x1 + Omega3 x4

Spleen rupture = Hydrolysed Collagen +
        Flaxseed oil
Flaxseed oil loves
    Haemorrhoids

Uva Ursi + Aminoi500
= Bladder stability

Brainstem = Ashwagandha +
        Omega 3 oils

Harry Potter

3.32pm mesoface x1 + Uva x1 + aminol5oo x1 + watery left eye

4.37pm: mouth sore gone + flaxseed x4

6.47pm: Uva x1 + silica x1 + lysine x1 + hydro collag x2 + arginine x2

8.28pm: flaxseed x4 + silica x1

9.12pm: saw palm x1 + hydro collag x2 + aminol5oo x1

9.41pm: omega3 x4

10.20pm: hydro collag x2 + lysine x1

February 24, 2019

Spinal cord (Injury Pain, disorder)
vertebrae
dislocation
— Pumpkin oil for (brain stem)
medulla oblongata (krill oil)

— tissues
amino 1500 +
high dose
glucosamine
sulphate +
cod live
oil

Strengthen
vertebrae
with silica
complex,
Omega 3 high dose
garlic + 5HTP
Ashwagandha

tissues (moisturises)
= Evening Primrose
oil
+ marine collagen +
Hydrolysed collagen

Paraplegia Paralysis (lower back)

L-Arginine (High dose)
Evening Primrose oil
Ashwagandha

10:55pm: ashwag x2 + codliver x1

11:49pm: flaxseed x4

FEBRUARY 25 2019 mJ 116th month

2:25am: flaxseed x4 + uva x1 + silica x1 + codliver x1 + arginine x2 + aminoisoa x1 + lysine x1

2:45am: ashwag x1

5:32am: flaxseed x4 + Hydrocollag x2

5:45am: moisture ankle bone intensity + Ever prim oil x1

8:30am: menopace x1 + omega3 x4

8:56am: uva ursi x1

9:14am: omega3 x4

Harry Potter

Ankle Tightness
Reduction = Omega 3 +
Amino 1500 + Flaxseed oil
+ Menopace + Marine Collagen
+ Hydrolysed Collagen
+ Brewers Yeast + Chromium

11:20 am : Omega3 × 4

12:06 pm : Omega3 × 4
about 1 pm : Flaxseed × 4
2:36 pm : Omega3 × 4 + UVA × 1
+ amino1500 × 1 + l-arginine × 2
+ myplace × 1
    ankle bone contractions
2:54 pm : ashwag × 1 + tan × 1

3:44 pm : l-arginine × 3 right mid
back (new) pain + Omega3 × 4

5:04 pm : Omega3 × 4

5:43 pm : hydro collag × 3

6:32 pm : garlic × 3 + Omega3 × 4
+ saw palm × 2

11:20 am: Omega3 × 4

12:06 pm: Omega3 × 4
about 1 pm: Flaxseed × 4
2:36 pm: Omega3 × 4 + uva × 1
+ amino1500 × 1 + l-arginine × 2
+ magpotace × 1
      ankle bone contractions
2:54 pm: ashwag × 1 + tan × 1

3:44 pm: l-arginine × 3 right ^{side} mid
back (new) pain + omega3 × 4

5:04 pm: Omega3 × 4

5:43 pm: hydro collag × 3

6:32 pm: garlic × 3 + Omega3 × 4
+ saw palm × 2

8pm: amino 1500 × 3
Some foot pinches
changed shoe to Cluggs or pink
scholl

8:32pm: face wash + Some tPtal +
chromium × 1 + silica × 1 + omega3 × 4

9:17pm: ashwag × 1

12:04am: Feb 2 6 2019
flaxseed × 3 + l-arginine × 2 + amino
1500 × 2 + Brewers yeast × 4

12:27am: menopace × 1

3:23 am: Omega3 × 5

3:33am: urq u3i × 1

Harry
Potter

7.53am: omega3 x 5 + urax 1
+ amino15 cox 1

8.34am: flaxseed x 3
+ hydro collag x 2

home by 2.33pm
gas, elec, water → bus 266 → kaustel
→ Sainsburys breadsticks, malt, donuts,
sauce, babyoil, fishray → bus 440 → DHL
Disability letter about £8
→ bus 440 → momsons + prescription →
bus 266 → Hammersmith Boots Tubular
Support Bandage + M&S Brewers yeast + garlic
→ bus 266 → home
3.14pm: garlic x 6   Bandage on
                        feature shoes
                        out
3.36pm: Brewers yeast x 6

4.15pm: Brewers yeast x 8

5.07pm: Brewers yeast x 8

Harry
Potter

Omega3 oil on High Dose
Peels off the black
 impurities of the skin
  follicles and cleanses
out dark impurities.

L-Arginine grows the hair
out with Marine collagen +
Saw palmetto

Harry Potter

6.25pm: Brewer yeast x8

7.03pm: SMTP x1

8.24pm: Brewer yeast x8
+ flaxseed x3 + ura x1 + ashwag x1

9.43pm: Omega 3 x5

10.28pm. ashwag *3 +
back wash + irritation +
Brewer yeast x8 + omega 3 x 5
+ arginine x3

11.12pm: milk thistle x1

1.29am Feb 27 2019
Omega 3 x5

Harry Potter

1.55am: flaxseed x3 + ashwag x1
+ Brewers yeast x7

6.07am: flaxseed x3 + uva x1
+ Brewers yeast x8 + omega3 x6

9am: Omega3 x5 + uva x1
+ Brewers yeast x6

9.20am: Hydrolysed Collagen x2

home by 2.07pm
   bus 440 → 2.07/427 →
AJ Nature Ashwag fulka less (30)
↳ bus 427/207 → Morrisons →
bus 266 → Hammersmith →
Iceland → £35.10 home delivery
7 → 9pm → bus 266
→ traffic → home
sneezed and sniffed all
   the way

2.35pm: l-arginine x3

4.40pm: l-arginine x3 +
Brewers yeast x8
sneezed over twelve times
4.45pm: ashwag x1.
sneezed over + sniffing
4.54pm: flaxseed x3

5.13pm: Brewers yeast x8

5.26pm: ashwag x1 + amino1500 x2
+ uva x1 sniffing

6.21pm: amino 1500 x3

6.53pm: l-arginine x3

7.16pm: amino1500 x3 +
ashwag x1

7.49pm: flaxseed x3

8.19pm: sniffing + omega3 x2

8.56pm: some sniffing omega3 x2

9.21pm: omega3 x2

Iceland home delivery

10.21pm: toptail + ashwag x1
+ l-arginine x3 + garlic x6
itchy back on oil

10.53pm: Chromium pico x1

11.31pm: menopace x1 + tar x1
+ charcoal x1

11:50pm: garlic x6 + lysine x1 + Hydro collag x1

Feb 28 2019   1:37am:

Silica x1

3:47am: Brewer's yeast x6
+ amino 1500 x1 + silica x1
whippy feel on dorsal foot front
+ upper right tooth pain

4:02am: l-arginine x3 foot arch
'smaller' feel left

4:24am: amino 1500 x3
+ omega3 x4 + t aur x1
+ ashwag x1 + Ura x1
achilles tendon some pain

11:03am: omega3 x5 + uraurix1 + ashwag x1

11:30am: Brewers yeast x7 + l-argnine x2

12:08pm: Omega3 x5 + taur x1

12:13pm: ashwag x1

2:05pm: arginine x2 + Omega3 x5 Brewers yeast x5 + Chromium x1

3:03pm: urax1 + arginine x2 + Brewers yeast x6 + ashwag x1

4:02pm: Silica x1

Harry Potter

5.05pm: flaxseed x2 + amino1500 x2

5.51pm: arginine x3 + omega3 x4
+ uraursi x1 + saw palm x2

6.51pm. chromium x1

8.24pm. garlic x7

9.28pm. Brewers yeast x7

10.15pm: garlic x2 + omega3 x4
+ arginine x2 + chromium x1
+ valerian x2

11.34pm. silica x1 + omega3 x4
+ tart x1

12.26am ~~Feb~~ March 1 2019
toptail shower with exfolia
+ relaxing and sweet
foot still swollen but walked to room
no shoe

12:26am March 1 2019 4 yrs today
Liverx1

+ amino1500x3 + ashwag x1 +
hydro collag x2 + Omega3 x4

12:54 am: Uvaursix1 + flaxseedx1

3:58 am: flaxseedx1 + uvax1
+ Brewers yeast x6 + amino1500x3
+ hydro collagx1

4:30 am: silicax1 + Omega3 x4

5:26 am: flaxseedx3 itchy
irritated skin

8am: watery left eye + eveprimx1
+ arginine x2 + silicax1 + flaxseedx3

10:47am: flaxseed x3 + arg inne x1
+ even prim oil x1

11:12 am: flaxseed x4 +
aminol 500 x1

11:48am: flaxseed x3 +
pumping excess fat out of
my eyes causing teary eyes

12.23 pm: aminol saw x2 +
omega3 x3

3:17pm: flaxseed x2 + saw palm x2

5:37pm: Brewers yeast x6 +
flaxseed x2

6.06pm: Brewers yeast x6

Harry Potter

Lower Back Health ⟹
   Flaxseed + Brewers yeast

march 1
2019   INVENTION
   Toptail shower Bowl with
rollers and stoppers and towel railings
Small, medium and large sizes
Toptail standing shower bowl

6.53pm: Cranbery x3 + Saw palm x2
+ Brewer yeast + x6 + amino 1500 x1
+ flaxseed x2

7.22pm: Omega3 x3
+ some sit-on-bed yoga

8.11pm: Silica x1

8.25pm: Chromium x1

8.35pm: menopace x1 + tan x1

9.35pm: Omega3 x4 + front tooth
right pain linked to nasal pain

9.40pm: amino 1500 x3 + arginine x1
+ ashwag x1

9.45pm: Brewer yeast x6

Harry
Potter

10.33pm: Omega 3 x5 + flaxseed x 2
    Painful teeth
10.44pm: even pain x 2
    stiff teve teeth

11.02pm: Chromium x1 + ashwag x1
+ Brewes yeast x 6 + intense
teeth nasal pain migraine

1.33 am [March 2 2019]

Brewes yeast x6 + l-arginine x 3
+ chromium x1

7.37 am: Brewes yeast x 6 +
l-arginine x 3 + tan x 1
+ chromium x 1

Harry Potter

grow your Hair

Saw Palmetto + L-Arginine +
Marine collgen

You must always seek
Doctor's advice before taking
ANY medicine!

Harry
Potter

11.44am: l-arginine×3 + Chromium×1
+ Brewers yeast×6 + flaxseed×2

12.44pm: Brewers yeast×6

2.51pm: Chromium×1 + Omega3×5
+ ashwag×1 + arginine×3

4.14pm: amino15co×2

4.23pm: nasal front teeth (right)
migrane + garlic (2008mg) 6
salt in water for painful teeth

4.35pm: flaxseed×3
Period pain

5.17pm: garlic×6 +
Omega3×4

5.52pm: Omega3 x4

6.12pm: Omega3 x4 nasal truth
eye migraine

6.46pm: Omega3 x4

7.43pm: Omega3 x4 + flaxseed x2

8.20pm: Omega3 x4 + garlic x6

8.48pm: Omega3 x4

9.40pm: Omega3 x4 + ashwag x1
+ flaxseed x2 + cranberry x6

11.31pm: L-arginine x2

Harry
Potter

2:04am 〔march 3 2019〕
Omega3x4 + garlicx6

2:20am: l-argininex1

8:03am: omega3x4

8:21am: arginnex1 + flaxseedx2
+ garlicx6

11:54am: sneeze1 + uvax1
+ hydro collagx3

12:36pm: worn out
+ arginnex1 + even primoilx1
+ omega3x4

2:19pm: l-argininex1
wrinkled eyes

Harry Potter

4.30pm: arginine x1 + Omega3 x4

5.34pm: alpha lipoic x1 + uva x1
+ arginine x1 + aminoISO x1

6.59pm: hemp seed oil x1

8.24pm: Omega3 x4 + hempseed x1

9.52pm: hemp oil x1 + aminoISO x1
+ omega3 x4 + flaxseed x1
+ menopace x1 hand pain, some
shoulder pain, period pain

10.51pm: ashwag x1 + arginine x1
+ exfoliation + toptail shower +
hempseed x1 + lysine x1 + charcoal x1
+ cleansed shoe + omega3 x4 +
flaxseed x2 + aminoISO x1

Foot Arch loves Omega 3
On High Dose

Teeth loves Omega 3 (High Dose)

2:46 am | march 4 2019 |
hemp seed x1 + saw palm x1 +
flaxseed x2 + omega 3 x 3+ arginnex

9:47 am : amino 1500 x 3
+ tan x1 Stiff right clavicle +
right side waist between front & Back
pain

11:27 am : amino 1500 x 3 + arginnex

www.ingramcontent.com/pod-product-compliance
Lightning Source LLC
Chambersburg PA
CBHW070947200526
45161CB00001BA/24